I0756098

THIS BOOK BELONGS TO:

THE WONDERFUL
WORLD of OWLS

Dedicated to all who love owls.

ISBN 978-1-970416-06-0

www.joeysavestheday.com

Mimi Books™ Publishing

A Mimi Book

Owls live on every continent except Antarctica. They can be found in forests, deserts, grasslands, and even cities.

Owls are birds, but unlike most birds, they are active at night. This makes them nocturnal, meaning they sleep during the day and hunt after dark.

There are over 250 species of owls around the world. Each species has its own size, color, and special adaptations.

Owl eyes are huge compared to their heads. Their big eyes help them see in very low light. They can't move their eyes side to side. Instead, their eyes are fixed in place like binoculars.

Owl eyes come in different colors: yellow, orange, or dark brown. The color often hints at when they prefer to hunt.

270°

Owls can turn their heads up to 270 degrees. This is almost all the way around! They can do this because they have extra bones in their necks. Humans have 7 neck bones, but owls have 14.

14 Neck bones

Owls have incredible hearing that helps them hunt in total darkness.

THE *Hunt* IS ON

Many owls have one ear higher than the other. This uneven placement helps them pinpoint exactly where sounds come from.

Their feathers are soft and velvety to the touch. This texture absorbs sound. Even when flapping hard, owls make almost no noise.

Owls are carnivores, meaning they eat meat.

CARNIVORES

Their favorite foods include mice, voles, insects, and small birds.

Large owls can hunt rabbits, skunks, and even small foxes.

Owl Pellets

Owls swallow small prey whole. Later, they cough up pellets made of bones and fur. Pellets help scientists learn what owls eat and where they hunt.

Cute!

Baby owls are called owlets.
They are fluffy, round, and very cute.

Protect!

Owl parents take turns protecting the nest. One hunts while the other stays with the babies.

Owlets learn to fly by hopping from branch to branch. This stage is called “branching.”
Learning

Growing

Young owls stay with their parents until they can hunt on their own.

Tree Hollow

Owls don't build their own nests. They use tree hollows, old nests, barns, or rocky cliffs.

Barn owls often nest in barns, attics, or abandoned buildings.
Barn Owls

Burrowing Owls

Burrowing owls live underground in tunnels made by prairie dogs.

Snowy owls nest on the ground in the Arctic tundra.

Snowy Owl

The barn owl is known for its heart-shaped face and ghostly white feathers.

Great horned owls have feather tufts that look like ears. They are powerful hunters.

Snowy owls are bright white and live in cold Arctic regions.

Bright

Elf owls are the smallest owl species. They are tiny enough to fit in your hand.

The Eurasian eagle-owl is one of the largest owls in the world.

Eurasian Eagle-Owl

Some owls hoot to communicate with each other. Each species has its own call.

Hoot!

Wild

Owls can live up to 25 years in the wild.

In North America, the great horned owl is one of the most common species. In Europe, the tawny owl is well known for its "too-wit too-woo" call.

North America

Europe

In Australia, the powerful owl is the largest owl species. In Asia, the scops owl is famous for its tiny size and big voice.

In Africa, the pearl-spotted owlet is active during the day.
Africa

Protect

People can help owls by protecting forests and putting up nest boxes.

Count the owls.

Thank you for exploring The Wonderful World of Owls with me. I hope you learned something new and enjoyed discovering these amazing, wide-eyed nighttime birds.

If you liked this book, please consider leaving a review. It helps other families discover it too.

See you in the next adventure!

Check out these other interesting books in the Wonderful World of series!

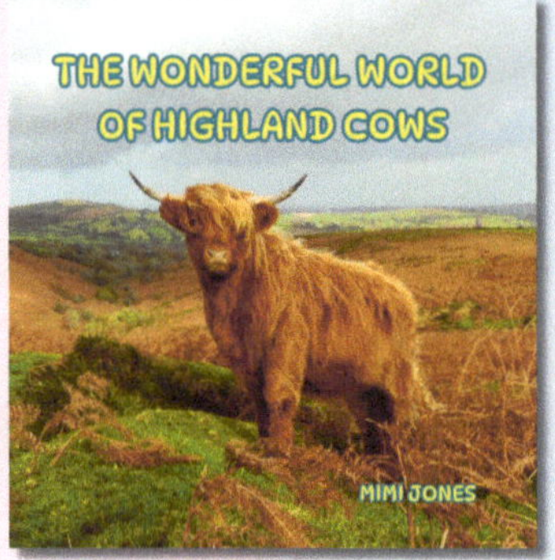

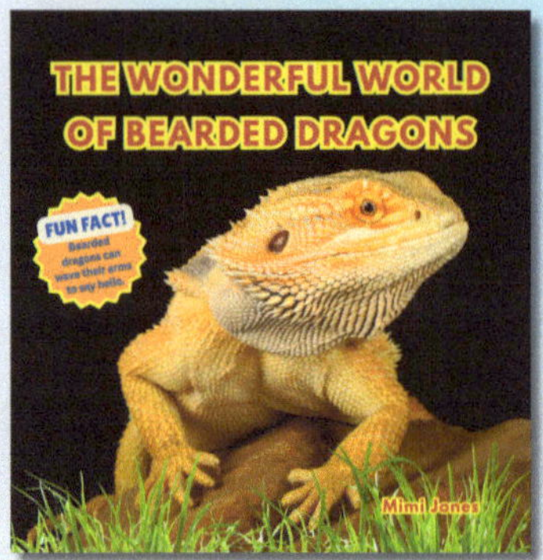

www.mimibooks.com

www.ingramcontent.com/pod-product-compliance
Lightning Source LLC
LaVergne TN
LVHW070158110826
845147LV00002B/437

* 9 7 8 1 9 7 0 4 1 6 0 6 0 *